I0797701

ANTS

by Martha London

Cody Koala
An Imprint of Pop!
popbooksonline.com

abdobooks.com
Published by Pop!, a division of ABDO, PO Box 398166, Minneapolis, Minnesota 55439.

Printed in the United States of America, North Mankato, Minnesota

082020
012021

THIS BOOK CONTAINS RECYCLED MATERIALS

Cover Photo: iStockphoto
Interior Photos: iStockphoto, 1, 5 (top), 5 (bottom left), 7, 9, 10, 19 (top), 19 (bottom left), 19 (bottom right), 21; Shutterstock Images, 5 (bottom right), 13, 14, 17

Editor: Christine Ha and Brienna Rossiter
Series Designer: Sophie Geister-Jones

Library of Congress Control Number: 2019954997

Publisher's Cataloging-in-Publication Data
Names: London, Martha, author.
Title: Ants / by Martha London
Description: Minneapolis, Minnesota : POP!, 2021 | Series: Underground animals | Includes online resources and index.
Identifiers: ISBN 9781532167584 (lib. bdg.) | ISBN 9781532168680 (ebook)
Subjects: LCSH: Ants--Juvenile literature. | Insects--Juvenile literature. | Ants--Behavior--Juvenile literature. | Burrowing animals--Juvenile literature. | Underground areas--Juvenile literature.
Classification: DDC 595.79/6--dc23

Hello! My name is

Cody Koala

Pop open this book and you'll find QR codes like this one, loaded with information, so you can learn even more!

Scan this code* and others like it while you read, or visit the website below to make this book pop.

popbooksonline.com/ants

*Scanning QR codes requires a web-enabled smart device with a QR code reader app and a camera.

Table of Contents

Chapter 1

Making Nests

Ants are insects. They live in large groups called **colonies**. A colony can have millions of ants. The ants live together in a nest.

Watch a video here!

Many ants build their nests underground. Nests have lots of tunnels. One tunnel leads to the surface. Other tunnels go between rooms. Ants use the rooms to store food or raise young.

Some ants can chew through wood. They make their nests in buildings or trees.

tunnel

Chapter 2

Parts of an Ant

An ant's body has three sections. They are the head, **thorax**, and **abdomen**. An ant also has six legs.

Learn more here!

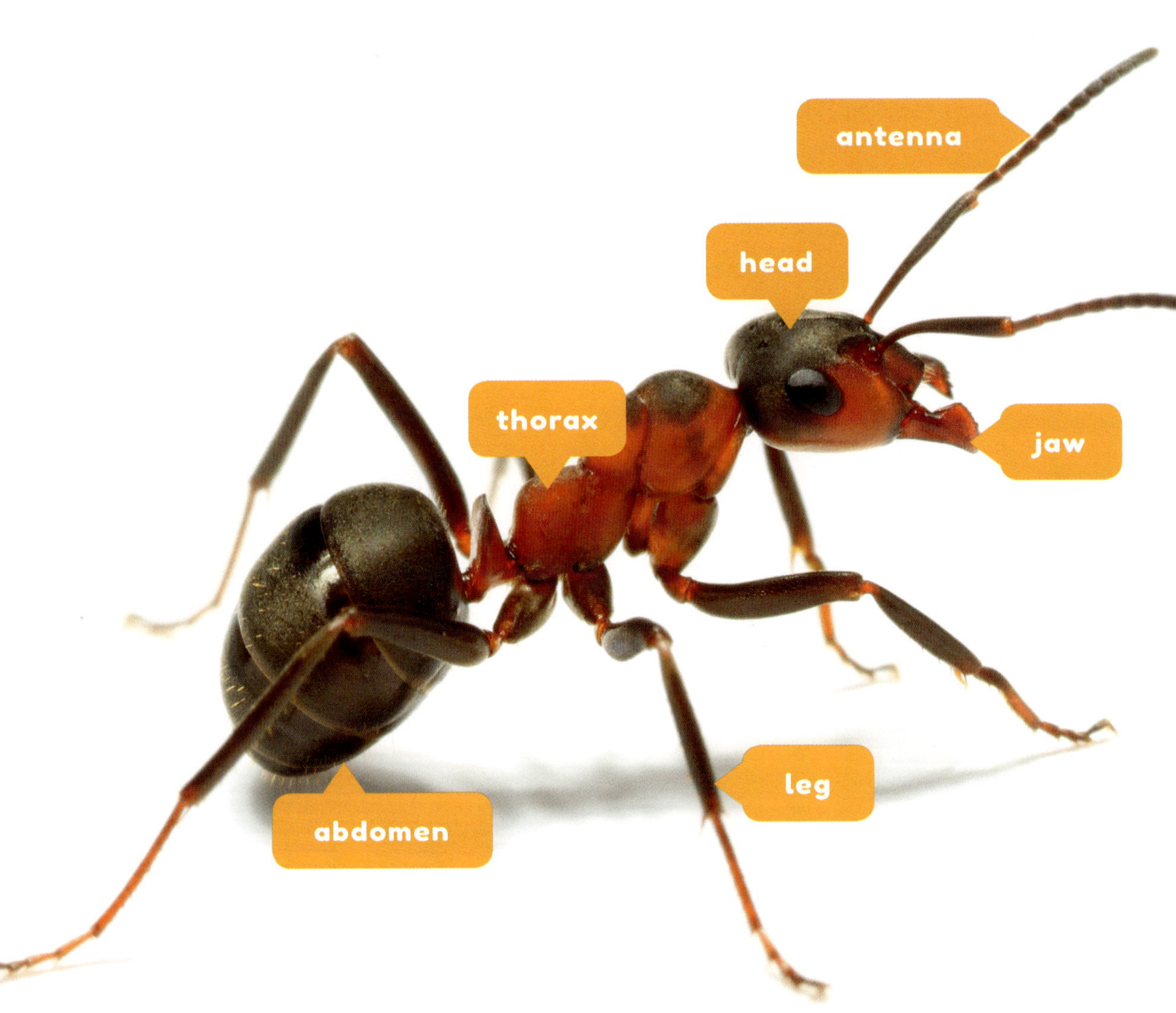
antenna
head
jaw
thorax
abdomen
leg

Ants have two antennae. The antennae help ants sense their surroundings.

Ants have strong jaws too. They use their jaws to dig. The jaws can also carry food.

Chapter 3

Life in a Colony

Three types of ants live in every **colony**. Each type of ant has a different job. A colony usually has one queen. The queen ant lays eggs.

eggs
Complete an activity here!

Male ants are called drones. They **mate** with the queen. The rest of the ants are workers. Worker ants help build the nest. They collect food. They also care for the queen's eggs.

All worker ants are female.

Larvae hatch from the eggs. Workers feed the larvae. The larvae grow. They form hard shells around their bodies. Adult ants come out.

larva

Chapter 4

Working Together

The nest keeps ants safe from **predators**. Worker ants leave the nest to find food. Ants eat plants, seeds, and some insects. Some ants drink **nectar**.

Learn more here!

Workers bring the food back to the nest. They often carry the food on their backs. Sometimes, several ants help lift one piece of food.

Ants can lift 20 times their body weight.

Making Connections

Text-to-Self

Have you ever seen a colony of ants? Where was it?

Text-to-Text

What books have you read about other insects? How are ants similar to the insects in those books? How are ants different?

Text-to-World

What other animals work together to get food?

Glossary

abdomen – the hind part of an insect's body.

colony – a group of animals that live together.

larva – the first life stage of an insect, often looking like a worm.

mate – to come together to have babies.

nectar – a sweet, sugary liquid made by a plant.

predator – an animal that hunts other animals for food.

thorax – the middle part of an insect's body.

Index

Online Resources

popbooksonline.com

Thanks for reading this Cody Koala book!

Scan this code* and others like it in this book, or visit the website below to make this book pop!

popbooksonline.com/ants

*Scanning QR codes requires a web-enabled smart device with a QR code reader app and a camera.